DOUBLE DIGIT MATH
PRACTICE WORKBOOK
SIMPLE ADDITION AND SUBTRACTION

{ **THIS BELONGS TO:** }

_____

NAME: _____ **DATE:** _____

| tens | ones | | tens | ones | | tens | ones | | tens | ones | | tens | ones |

	1	7
+	0	2

	2	7
+	3	1

	2	4
+	0	3

	5	4
+	1	0

	2	3
+	2	6

	5	5
+	4	2

	1	4
+	2	3

	3	3
+	2	6

	8	6
+	1	2

	4	3
+	1	3

	1	3
+	7	2

	3	4
+	1	0

	2	2
+	3	7

	1	2
+	4	5

	1	7
+	8	2

	6	7
+	1	2

	1	3
+	5	3

	3	4
+	4	2

	7	8
+	2	1

	0	7
+	9	1

	1	7
+	3	2

	4	1
+	3	8

	6	7
+	2	2

	1	4
+	3	2

	1	8
+	2	1

1

tens	ones
3	4
+ 1	3

tens	ones
1	2
+ 1	5

tens	ones
3	5
+ 0	3

tens	ones
1	4
+ 7	0

tens	ones
2	3
+ 4	6

tens	ones
2	5
+ 1	2

tens	ones
1	6
+ 4	3

tens	ones
3	2
+ 5	5

tens	ones
7	6
+ 1	2

tens	ones
4	3
+ 2	4

tens	ones
0	3
+ 9	2

tens	ones
3	4
+ 0	0

tens	ones
1	2
+ 5	7

tens	ones
2	2
+ 6	7

tens	ones
1	3
+ 7	2

tens	ones
4	4
+ 1	2

tens	ones
1	3
+ 2	6

tens	ones
7	4
+ 1	3

tens	ones
2	8
+ 2	1

tens	ones
0	5
+ 9	1

tens	ones
2	5
+ 4	3

tens	ones
0	4
+ 2	2

tens	ones
1	4
+ 8	1

tens	ones
1	7
+ 0	2

tens	ones
1	6
+ 3	2

NAME: _____ **DATE:** _____

tens	ones		tens	ones		tens	ones		tens	ones		tens	ones
	1	3			4	7			2	4			4	4			3	3
+	3	2		+	3	1		+	2	3		+	1	0		+	1	6

	5	9			1	0			1	3			3	6			6	3
+	4	0		+	2	3		+	1	6		+	2	2		+	1	3

	1	3			3	2			2	2			1	2			0	7
+	5	2		+	1	0		+	3	1		+	8	5		+	0	2

	6	7			2	0			3	4			5	8			1	7
+	1	1		+	5	3		+	5	2		+	2	0		+	3	1

	1	7			1	8			1	5			3	3			2	7
+	9	2		+	9	8		+	5	2		+	9	2		+	1	2

3

NAME: _____ DATE: _____

tens	ones		tens	ones		tens	ones		tens	ones		tens	ones

	1	3
+	3	2

	4	7
+	4	1

	5	2
+	0	3

	6	4
+	1	0

	1	3
+	2	1

	2	6
+	4	2

	1	4
+	0	3

	4	3
+	1	6

	8	9
+	1	0

	0	8
+	1	1

	1	3
+	1	2

	3	4
+	1	2

	2	2
+	3	0

	1	2
+	5	5

	4	7
+	3	2

	6	1
+	1	3

	1	5
+	3	4

	2	4
+	4	2

	2	4
+	3	3

	0	8
+	2	1

	1	4
+	7	2

	0	4
+	3	2

	1	6
+	7	2

	3	7
+	2	2

	1	7
+	0	2

NAME: _____ **DATE:** _____

tens	ones
1	7
+ 5	2

tens	ones
2	3
+ 3	4

tens	ones
2	4
+ 1	3

tens	ones
5	4
+ 1	1

tens	ones
2	3
+ 0	6

tens	ones
5	5
+ 4	4

tens	ones
2	4
+ 2	3

tens	ones
3	3
+ 2	0

tens	ones
0	6
+ 1	2

tens	ones
4	3
+ 2	3

tens	ones
1	3
+ 9	2

tens	ones
3	4
+ 4	0

tens	ones
2	2
+ 3	3

tens	ones
1	2
+ 3	5

tens	ones
1	7
+ 8	1

tens	ones
3	7
+ 5	2

tens	ones
4	3
+ 3	2

tens	ones
2	4
+ 4	5

tens	ones
3	8
+ 2	1

tens	ones
0	8
+ 8	0

tens	ones
0	7
+ 8	2

tens	ones
2	7
+ 2	2

tens	ones
1	1
+ 3	3

tens	ones
6	0
+ 5	2

tens	ones
3	7
+ 9	2

NAME: _____ DATE: _____

tens	ones
1	7
− 0	2

tens	ones
4	7
− 3	1

tens	ones
5	4
− 1	3

tens	ones
6	4
− 2	0

tens	ones
4	5
− 2	1

tens	ones
5	6
− 1	2

tens	ones
4	4
− 2	3

tens	ones
9	8
− 2	6

tens	ones
2	6
− 1	2

tens	ones
4	5
− 3	3

tens	ones
4	9
− 1	2

tens	ones
5	6
− 2	0

tens	ones
5	5
− 1	0

tens	ones
9	4
− 4	1

tens	ones
5	7
− 2	2

tens	ones
1	7
− 1	2

tens	ones
5	9
− 5	3

tens	ones
4	4
− 1	2

tens	ones
5	8
− 2	1

tens	ones
9	5
− 0	2

tens	ones
4	6
− 2	1

tens	ones
6	7
− 3	2

tens	ones
4	8
− 1	5

tens	ones
8	9
− 5	2

tens	ones
1	7
− 0	4

NAME: _____ DATE: _____

| tens | ones | | tens | ones | | tens | ones | | tens | ones | | tens | ones |

		2	7
−	0	2

	2	7
−	1	1

	2	4
−	0	3

	5	4
−	1	0

	4	8
−	2	6

	5	5
−	4	2

	1	4
−	0	3

	3	6
−	2	6

	8	6
−	1	2

	4	3
−	1	3

	9	6
−	7	2

	3	4
−	1	0

	5	7
−	3	7

	6	6
−	4	5

	6	7
−	5	3

	4	6
−	2	2

	5	7
−	5	3

	5	6
−	4	2

	5	7
−	3	2

	5	6
−	2	1

	7	8
−	2	2

	5	6
−	4	2

	6	7
−	1	3

	9	8
−	9	2

	8	6
−	1	2

7

NAME: _____ **DATE:** _____

tens	ones		tens	ones		tens	ones		tens	ones		tens	ones
1	7		4	7		2	4		4	4		5	6
− 0	2		− 3	2		− 0	3		− 1	0		− 2	6

tens	ones		tens	ones		tens	ones		tens	ones		tens	ones
4	6		3	4		3	8		3	5		3	2
− 4	2		− 2	0		− 2	6		− 1	2		− 1	2

tens	ones		tens	ones		tens	ones		tens	ones		tens	ones
4	4		5	4		5	7		6	5		9	7
− 2	2		− 1	0		− 3	7		− 4	5		− 8	1

tens	ones		tens	ones		tens	ones		tens	ones		tens	ones
6	7		6	4		5	4		7	8		0	7
− 4	3		− 5	3		− 4	2		− 2	4		− 0	1

tens	ones		tens	ones		tens	ones		tens	ones		tens	ones
1	7		9	7		8	3		5	7		1	7
− 0	2		− 9	2		− 3	2		− 2	2		− 0	2

NAME: _____ **DATE:** _____

tens	ones		tens	ones		tens	ones		tens	ones		tens	ones

	2	5
−	0	2

	4	4
−	3	1

	1	7
−	0	3

	5	5
−	1	0

	2	5
−	1	4

	6	5
−	2	2

	4	4
−	2	3

	3	7
−	2	6

	8	6
−	6	3

	5	3
−	4	3

	5	3
−	5	2

	3	4
−	0	0

	4	2
−	3	1

	5	2
−	4	0

	1	7
−	1	2

	4	2
−	1	2

	6	4
−	5	3

	5	4
−	4	2

	7	7
−	2	1

	0	7
−	0	1

	9	4
−	9	2

	1	7
−	1	2

	1	6
−	0	2

	6	5
−	3	2

	6	2
−	1	2

tens	ones		tens	ones		tens	ones		tens	ones		tens	ones
1	9		3	6		2	4		5	4		2	7
− 0	2		− 3	1		− 0	3		− 1	0		− 2	6

tens	ones		tens	ones		tens	ones		tens	ones		tens	ones
5	5		3	4		3	3		8	6		4	3
− 4	2		− 2	3		− 2	6		− 1	2		− 1	3

tens	ones		tens	ones		tens	ones		tens	ones		tens	ones
7	3		3	4		3	2		5	5		8	7
− 7	2		− 1	0		− 3	0		− 4	5		− 8	2

tens	ones		tens	ones		tens	ones		tens	ones		tens	ones
6	4		6	5		4	5		6	8		0	7
− 1	2		− 5	3		− 4	2		− 2	1		− 0	2

tens	ones		tens	ones		tens	ones		tens	ones		tens	ones
9	6		6	7		1	6		0	7		1	4
− 9	2		− 4	2		− 1	2		− 0	3		− 1	2

tens	ones
2	6
+ 0	2

tens	ones
2	5
+ 3	2

tens	ones
2	4
+ 0	3

tens	ones
4	4
+ 1	0

tens	ones
4	3
+ 2	6

tens	ones
5	5
+ 0	2

tens	ones
2	4
+ 2	3

tens	ones
6	3
+ 2	2

tens	ones
5	4
+ 1	2

tens	ones
4	2
+ 1	3

tens	ones
1	3
+ 2	2

tens	ones
3	3
+ 1	0

tens	ones
2	2
+ 2	5

tens	ones
1	2
+ 4	3

tens	ones
1	5
+ 5	2

tens	ones
3	7
+ 1	2

tens	ones
1	3
+ 2	3

tens	ones
2	4
+ 4	2

tens	ones
7	8
+ 2	0

tens	ones
0	7
+ 9	0

tens	ones
5	7
+ 4	1

tens	ones
9	6
+ 9	2

tens	ones
1	7
+ 9	2

tens	ones
5	6
+ 5	2

tens	ones
4	7
+ 3	3

| tens | ones | | tens | ones | | tens | ones | | tens | ones | | tens | ones |

	1	7
+	1	2

	2	8
+	3	1

	1	4
+	0	3

	5	2
+	1	0

	2	3
+	6	6

	5	5
+	4	3

	6	4
+	2	3

	3	3
+	2	6

	4	6
+	1	2

	4	3
+	1	3

	1	3
+	7	2

	3	4
+	1	0

	2	2
+	4	7

	7	2
+	4	5

	1	7
+	8	2

	6	6
+	1	2

	1	3
+	2	3

	6	4
+	4	2

	2	8
+	2	1

	1	7
+	9	1

	4	6
+	2	2

	6	7
+	3	3

	9	8
+	9	7

	5	3
+	5	2

	6	7
+	2	2

NAME: _____ DATE: _____

tens	ones		tens	ones		tens	ones		tens	ones		tens	ones

	1	7
+	2	2

	2	7
+	3	1

	2	4
+	0	0

	5	4
+	3	0

	2	3
+	1	6

	5	1
+	4	5

	1	4
+	4	3

	3	3
+	1	6

	9	6
+	1	2

	4	3
+	4	3

	1	3
+	8	2

	3	4
+	1	3

	3	2
+	3	7

	1	2
+	4	2

	1	7
+	5	2

	2	7
+	1	2

	2	3
+	5	3

	5	4
+	4	2

	7	8
+	2	1

	0	7
+	9	1

	1	2
+	4	2

	1	7
+	1	2

	1	4
+	3	2

	1	7
+	5	2

	1	6
+	9	2

13

NAME: _____ DATE: _____

tens	ones		tens	ones		tens	ones		tens	ones		tens	ones

	1	7			2	7			2	4			5	4			2	3
+	4	2		+	3	2		+	0	3		+	1	2		+	2	6

	5	5			1	4			3	3			5	6			4	3
+	4	2		+	2	3		+	2	6		+	1	2		+	1	4

	1	3			3	4			2	2			1	2			1	7
+	7	5		+	6	0		+	6	7		+	4	5		+	2	2

	6	7			1	3			3	4			7	5			0	7
+	2	2		+	8	3		+	5	2		+	2	1		+	5	1

	1	4			1	7			4	7			1	7			1	5
+	9	2		+	3	2		+	9	2		+	9	1		+	9	2

14

tens	ones		tens	ones		tens	ones		tens	ones		tens	ones

| | 1 | 6 | | | 2 | 7 | | | 2 | 6 | | | 5 | 4 | | | 5 | 3 |
| + | 2 | 2 | | + | 4 | 2 | | + | 2 | 3 | | + | 1 | 3 | | + | 2 | 6 |

| | 5 | 3 | | | 1 | 4 | | | 3 | 3 | | | 8 | 3 | | | 4 | 3 |
| + | 3 | 2 | | + | 1 | 3 | | + | 5 | 6 | | + | 1 | 4 | | + | 2 | 5 |

| | 1 | 7 | | | 3 | 4 | | | 2 | 2 | | | 1 | 2 | | | 1 | 7 |
| + | 7 | 2 | | + | 1 | 3 | | + | 4 | 7 | | + | 3 | 6 | | + | 5 | 1 |

| | 6 | 1 | | | 1 | 3 | | | 3 | 5 | | | 7 | 8 | | | 0 | 7 |
| + | 1 | 6 | | + | 4 | 4 | | + | 4 | 4 | | + | 1 | 1 | | + | 8 | 1 |

| | 1 | 3 | | | 1 | 5 | | | 1 | 5 | | | 1 | 6 | | | 1 | 9 |
| + | 5 | 3 | | + | 6 | 2 | | + | 3 | 2 | | + | 5 | 3 | | + | 5 | 0 |

NAME: _____ **DATE:** _____

tens | ones	tens | ones	tens | ones	tens | ones	tens | ones

	4	7			3	8			2	6			5	5			2	3
−	1	2		−	3	1		−	2	3		−	1	5		−	2	0

	5	5			6	7			3	9			8	6			9	3
−	2	2		−	2	3		−	2	6		−	1	2		−	1	3

	9	3			3	4			3	7			5	6			5	6
−	7	2		−	1	2		−	2	7		−	4	5		−	4	2

	5	8			7	5			5	4			7	8			7	6
−	1	2		−	5	3		−	4	2		−	2	4		−	3	1

	9	5			5	6			6	4			9	7			6	6
−	8	2		−	3	2		−	4	2		−	6	2		−	0	2

NAME: _____   DATE: _____

tens	ones		tens	ones		tens	ones		tens	ones		tens	ones

tens	ones
8	8
− 2	2

tens	ones
6	6
− 3	1

tens	ones
2	6
− 0	3

tens	ones
5	3
− 1	0

tens	ones
2	5
− 1	5

8	4
− 4	2

5	4
− 2	3

6	7
− 2	6

8	6
− 1	4

4	9
− 1	0

5	5
− 4	2

4	6
− 1	0

5	7
− 3	7

5	7
− 4	5

6	7
− 4	2

6	7
− 5	2

6	8
− 5	3

7	6
− 4	2

7	8
− 2	5

9	7
− 5	2

7	6
− 4	2

7	8
− 5	2

5	8
− 4	2

6	7
− 1	2

9	8
− 4	2

NAME: _____ DATE: _____

tens	ones		tens	ones		tens	ones		tens	ones		tens	ones

	1	5
−	0	2

	3	5
−	3	1

	6	6
−	0	3

	5	6
−	1	0

	6	6
−	2	4

	6	5
−	4	2

	2	5
−	2	3

	3	6
−	2	6

	7	7
−	1	2

	4	5
−	1	3

	7	6
−	7	2

	6	5
−	1	0

	8	7
−	3	6

	4	7
−	4	5

	8	7
−	8	5

	6	7
−	1	6

	5	5
−	5	3

	5	5
−	4	2

	6	8
−	2	1

	9	7
−	5	1

	9	7
−	9	0

	6	7
−	1	3

	8	7
−	5	5

	7	7
−	5	2

	9	9
−	1	2

18

NAME: _____   **DATE:** _____

tens	ones		tens	ones		tens	ones		tens	ones		tens	ones

Row 1:
- 9 8 / − 0 2
- 8 7 / − 3 1
- 7 7 / − 0 3
- 5 6 / − 3 0
- 2 6 / − 2 6

Row 2:
- 7 5 / − 4 2
- 9 4 / − 2 3
- 3 9 / − 2 6
- 7 6 / − 1 2
- 4 5 / − 1 3

Row 3:
- 7 5 / − 7 2
- 3 4 / − 1 4
- 8 8 / − 3 7
- 7 7 / − 4 5
- 9 9 / − 8 2

Row 4:
- 5 5 / − 1 2
- 6 3 / − 5 2
- 5 4 / − 4 2
- 9 8 / − 2 2
- 7 7 / − 5 1

Row 5:
- 7 7 / − 3 6
- 1 7 / − 0 2
- 9 7 / − 9 2
- 5 7 / − 5 2
- 6 7 / − 6 3

19

NAME: _____ **DATE:** _____

tens | ones	tens | ones	tens | ones	tens | ones	tens | ones

```
  1  2        6  7        4  4        5  5        2  7
- 0  2      - 3  1      - 0  3      - 1  0      - 1  6
_____    _____    _____    _____    _____


  5  4        4  4        3  6        8  5        4  4
- 4  2      - 2  3      - 2  6      - 5  2      - 1  3
_____    _____    _____    _____    _____


  5  3        3  4        9  6        6  8        7  7
- 2  2      - 1  2      - 3  2      - 4  5      - 1  2
_____    _____    _____    _____    _____


  6  5        6  3        9  4        9  8        9  7
- 5  2      - 5  3      - 4  2      - 2  1      - 9  1
_____    _____    _____    _____    _____


  6  8        5  9        4  7        3  5        5  7
- 5  2      - 4  2      - 2  4      - 2  2      - 4  2
_____    _____    _____    _____    _____
```

NAME: _____ DATE: _____

| tens | ones | | tens | ones | | tens | ones | | tens | ones | | tens | ones |

	1	5
+	0	2

	5	6
+	3	1

	3	7
+	0	2

	6	6
+	1	0

	5	7
+	2	1

	6	5
+	4	2

	6	4
+	3	3

	3	3
+	2	1

	8	8
+	1	1

	4	3
+	1	2

	9	9
+	5	0

	7	7
+	1	0

	6	5
+	3	4

	7	6
+	1	0

	9	9
+	8	0

	6	7
+	5	2

	1	3
+	5	4

	3	2
+	4	7

	7	8
+	2	1

	0	7
+	9	1

	1	7
+	4	2

	2	7
+	9	2

	1	7
+	7	2

	5	3
+	9	2

	1	5
+	9	2

21

tens	ones		tens	ones		tens	ones		tens	ones		tens	ones

	1	7
+	4	2

	2	7
+	4	2

	2	4
+	5	3

	5	4
+	1	2

	2	3
+	2	6

	5	5
+	4	4

	3	4
+	2	4

	7	3
+	2	6

	8	6
+	2	2

	4	3
+	1	5

	1	3
+	6	4

	3	4
+	1	4

	2	2
+	3	6

	1	2
+	5	5

	1	7
+	1	2

	6	7
+	1	0

	1	0
+	5	3

	3	6
+	4	2

	7	8
+	1	1

	0	7
+	8	1

	1	9
+	9	0

	1	7
+	0	2

	1	6
+	9	2

	5	7
+	0	2

	4	7
+	5	2

NAME: _____ DATE: _____

tens	ones		tens	ones		tens	ones		tens	ones		tens	ones

	8	7
+	0	2

	2	0
+	3	1

	3	4
+	0	3

	5	6
+	1	0

	2	9
+	2	0

	6	5
+	4	2

	1	4
+	0	3

	3	5
+	2	3

	8	6
+	2	2

	4	6
+	1	3

	1	3
+	5	2

	5	4
+	1	0

	6	2
+	3	7

	4	2
+	4	5

	6	7
+	2	2

	6	7
+	2	2

	6	4
+	5	3

	5	9
+	4	0

	6	7
+	2	1

	9	9
+	0	0

	1	7
+	0	2

	7	5
+	2	2

	5	5
+	2	2

	7	8
+	1	2

	5	9
+	2	2

23

NAME: _____ DATE: _____

	tens	ones
	1	7
+	0	0

	tens	ones
	2	7
+	9	1

	tens	ones
	2	4
+	6	3

	tens	ones
	5	7
+	1	0

	tens	ones
	2	3
+	0	6

	tens	ones
	5	5
+	4	4

	tens	ones
	1	4
+	2	2

	tens	ones
	3	2
+	2	7

	tens	ones
	8	6
+	1	3

	tens	ones
	4	3
+	1	2

	tens	ones
	8	6
+	7	2

	tens	ones
	3	5
+	1	0

	tens	ones
	2	6
+	3	4

	tens	ones
	1	5
+	4	2

	tens	ones
	1	7
+	9	2

	tens	ones
	6	6
+	1	2

	tens	ones
	1	3
+	6	4

	tens	ones
	3	4
+	4	5

	tens	ones
	7	8
+	3	1

	tens	ones
	0	6
+	9	1

	tens	ones
	1	4
+	9	2

	tens	ones
	1	2
+	2	2

	tens	ones
	1	8
+	9	1

	tens	ones
	1	7
+	5	2

	tens	ones
	1	5
+	5	3

tens	ones
1	7
+ 0	2

tens	ones
2	7
+ 3	1

tens	ones
2	4
+ 0	3

tens	ones
5	4
+ 1	0

tens	ones
2	3
+ 2	6

tens	ones
5	5
+ 4	2

tens	ones
1	4
+ 2	3

tens	ones
3	3
+ 2	6

tens	ones
8	6
+ 1	2

tens	ones
4	3
+ 1	3

tens	ones
1	3
+ 7	2

tens	ones
3	4
+ 1	0

tens	ones
2	2
+ 3	7

tens	ones
1	2
+ 4	5

tens	ones
1	7
+ 8	2

tens	ones
6	7
+ 1	2

tens	ones
1	3
+ 5	3

tens	ones
3	4
+ 4	2

tens	ones
7	8
+ 2	1

tens	ones
0	7
+ 9	1

tens	ones
5	6
+ 1	2

tens	ones
5	5
+ 9	2

tens	ones
1	7
+ 5	0

tens	ones
1	8
+ 9	0

tens	ones
1	6
+ 3	2

| tens | ones | | tens | ones | | tens | ones | | tens | ones | | tens | ones |

Row 1:
- 1 7 − 0 2
- 5 7 − 3 1
- 2 4 − 1 3
- 4 5 − 1 0
- 4 8 − 2 6

Row 2:
- 4 5 − 4 2
- 6 6 − 2 3
- 3 5 − 2 2
- 7 6 − 1 2
- 3 6 − 1 3

Row 3:
- 7 5 − 2 2
- 3 5 − 1 0
- 6 8 − 3 7
- 5 6 − 4 5
- 8 7 − 8 2

Row 4:
- 7 7 − 1 2
- 6 6 − 5 0
- 3 7 − 0 2
- 4 8 − 2 1
- 9 7 − 5 1

Row 5:
- 7 5 − 5 2
- 5 6 − 4 2
- 7 6 − 4 2
- 6 4 − 4 2
- 7 5 − 3 2

tens	ones		tens	ones		tens	ones		tens	ones		tens	ones

Row 1:
- 1 6 − 0 2
- 4 7 − 3 1
- 5 4 − 0 3
- 3 4 − 1 0
- 2 7 − 2 6

Row 2:
- 5 9 − 4 2
- 4 8 − 2 3
- 3 6 − 2 6
- 8 7 − 1 2
- 4 4 − 1 3

Row 3:
- 9 3 − 7 2
- 5 4 − 1 0
- 5 7 − 3 7
- 6 6 − 4 5
- 5 7 − 5 2

Row 4:
- 6 7 − 1 2
- 6 3 − 5 3
- 7 4 − 4 2
- 9 8 − 2 1
- 2 7 − 2 1

Row 5:
- 4 6 − 2 2
- 9 8 − 9 2
- 7 6 − 2 2
- 8 8 − 1 2
- 8 7 − 6 2

NAME: _____ DATE: _____

tens	ones		tens	ones		tens	ones		tens	ones		tens	ones

	5	7
−	0	2

	4	4
−	3	1

	3	6
−	0	3

	5	6
−	1	0

	3	6
−	2	6

	8	6
−	4	2

	4	4
−	2	3

	9	5
−	2	5

	8	9
−	1	2

	4	7
−	1	3

	8	5
−	7	2

	8	4
−	1	0

	9	2
−	3	7

	8	2
−	4	5

	9	7
−	8	2

	7	7
−	1	2

	6	4
−	5	3

	5	7
−	4	2

	6	8
−	2	1

	5	8
−	4	1

	8	6
−	5	2

	9	4
−	9	2

	4	6
−	2	2

	5	5
−	2	2

	9	7
−	9	3

28

tens	ones		tens	ones		tens	ones		tens	ones		tens	ones

| | 1 | 5 | | 6 | 6 | | 2 | 5 | | 5 | 7 | | 6 | 7 |
| − | 0 | 2 | − | 3 | 1 | − | 0 | 3 | − | 1 | 0 | − | 2 | 6 |

| | 5 | 5 | | 3 | 5 | | 3 | 3 | | 8 | 9 | | 4 | 8 |
| − | 1 | 2 | − | 2 | 3 | − | 2 | 2 | − | 1 | 2 | − | 1 | 3 |

| | 7 | 6 | | 3 | 5 | | 8 | 7 | | 5 | 5 | | 6 | 7 |
| − | 7 | 2 | − | 1 | 0 | − | 3 | 7 | − | 4 | 5 | − | 2 | 2 |

| | 8 | 5 | | 5 | 3 | | 7 | 4 | | 7 | 8 | | 9 | 7 |
| − | 1 | 2 | − | 5 | 0 | − | 4 | 4 | − | 2 | 4 | − | 2 | 1 |

| | 9 | 7 | | 1 | 7 | | 9 | 5 | | 5 | 6 | | 5 | 7 |
| − | 9 | 2 | − | 0 | 2 | − | 0 | 2 | − | 4 | 2 | − | 2 | 2 |

tens	ones		tens	ones		tens	ones		tens	ones		tens	ones

	1	7
−	0	2

	8	7
−	3	1

	2	8
−	0	3

	3	4
−	1	0

	2	6
−	2	6

	5	6
−	4	2

	2	8
−	2	3

	3	7
−	2	6

	7	8
−	1	2

	4	7
−	1	3

	7	8
−	7	2

	4	4
−	1	0

	5	5
−	3	3

	6	6
−	4	5

	9	7
−	8	3

	7	7
−	1	2

	5	6
−	5	3

	6	4
−	4	2

	6	6
−	2	1

	9	7
−	9	1

	9	7
−	9	0

	8	7
−	1	2

	6	8
−	6	2

	5	4
−	0	3

	5	5
−	4	2

30

NAME: _____ DATE: _____

| tens | ones | | tens | ones | | tens | ones | | tens | ones | | tens | ones |

```
    6   6          2   7          2   4          5   6          2   3
+   0   2      +   3   1      +   5   3      +   1   0      +   4   6
```

```
    5   7          6   6          5   4          5   6          4   6
+   3   2      +   2   3      +   2   2      +   1   2      +   1   3
```

```
    9   3          6   4          4   2          5   2          1   7
+   7   2      +   1   0      +   3   7      +   4   5      +   5   2
```

```
    6   7          1   3          3   3          7   5          8   7
+   2   2      +   5   2      +   4   2      +   2   1      +   9   1
```

```
    1   5          3   6          5   6          4   5          6   4
+   5   2      +   6   2      +   9   2      +   9   2      +   3   2
```

31

NAME: _____ DATE: _____

tens	ones		tens	ones		tens	ones		tens	ones		tens	ones
	3	7		9	7		2	4		5	5		2	3
+	3	2	+	3	1	+	4	3	+	2	0	+	3	6

	5	2		1	5		3	7		7	9		8	6
+	4	7	+	8	3	+	2	0	+	1	0	+	1	3

	2	3		3	4		2	6		1	5		1	8
+	7	5	+	1	3	+	3	2	+	4	4	+	6	1

	6	5		5	4		3	4		6	5		5	7
+	3	2	+	5	3	+	4	4	+	2	1	+	9	1

	1	6		1	3		1	7		1	7		0	3
+	4	2	+	3	5	+	5	2	+	0	2	+	9	2

32

NAME: _____   DATE: _____

tens	ones		tens	ones		tens	ones		tens	ones		tens	ones

	1	5
+	0	3

	5	6
+	3	1

	2	6
+	0	3

	5	6
+	1	2

	6	3
+	2	6

	5	5
+	4	4

	4	8
+	2	1

	5	9
+	2	0

	8	7
+	1	2

	6	3
+	1	3

	1	3
+	9	2

	3	4
+	1	4

	2	2
+	3	3

	5	3
+	4	5

	1	7
+	2	2

	6	7
+	1	0

	1	3
+	5	6

	3	4
+	8	2

	7	7
+	2	1

	0	3
+	3	1

	5	5
+	3	2

	8	8
+	7	2

	6	8
+	5	2

	9	6
+	9	2

	8	7
+	5	1

33

	tens	ones
	1	7
+	4	2

	tens	ones
	2	7
+	4	2

	tens	ones
	2	4
+	3	3

	tens	ones
	5	4
+	1	4

	tens	ones
	2	0
+	3	3

	tens	ones
	5	5
+	4	2

	tens	ones
	1	4
+	2	3

	tens	ones
	3	3
+	2	6

	tens	ones
	8	6
+	1	2

	tens	ones
	4	3
+	1	3

	tens	ones
	1	3
+	7	2

	tens	ones
	3	4
+	1	0

	tens	ones
	2	2
+	3	7

	tens	ones
	1	2
+	4	5

	tens	ones
	1	7
+	8	2

	tens	ones
	6	7
+	1	2

	tens	ones
	1	3
+	5	3

	tens	ones
	3	4
+	4	2

	tens	ones
	7	8
+	2	1

	tens	ones
	0	7
+	9	1

	tens	ones
	1	5
+	9	3

	tens	ones
	8	7
+	6	2

	tens	ones
	2	7
+	9	1

	tens	ones
	3	7
+	3	2

	tens	ones
	1	0
+	0	2

NAME: _____   **DATE:** _____

tens | ones	tens | ones	tens | ones	tens | ones	tens | ones

	1	7
+	2	2

	2	8
+	6	1

	2	6
+	5	3

	5	4
+	1	2

	2	3
+	5	6

	5	5
+	5	3

	1	5
+	4	3

	3	6
+	2	1

	8	7
+	1	2

	4	6
+	6	3

	1	3
+	8	4

	3	4
+	5	4

	2	2
+	0	7

	5	3
+	4	5

	3	3
+	8	2

	3	7
+	1	2

	1	1
+	5	8

	3	4
+	5	2

	7	8
+	2	0

	0	7
+	3	1

	1	7
+	7	1

	6	7
+	6	2

	5	6
+	2	1

	1	6
+	9	2

	5	7
+	5	2

tens	ones		tens	ones		tens	ones		tens	ones		tens	ones
5	5		2	7		9	4		6	4		9	6
− 0	1		− 0	1		− 0	3		− 1	0		− 2	6

tens	ones		tens	ones		tens	ones		tens	ones		tens	ones
5	5		2	6		3	7		8	8		4	7
− 4	2		− 2	3		− 2	6		− 1	2		− 1	3

8	3		3	4		2	2		6	8		9	7
− 7	2		− 2	2		− 0	0		− 4	5		− 8	2

6	7		3	4		3	4		5	5		6	6
− 0	2		− 2	3		− 3	3		− 2	1		− 2	1

9	7		5	5		6	7		5	7		6	6
− 1	3		− 0	2		− 6	3		− 3	3		− 1	2

NAME: _____ DATE: _____

| tens | ones | | tens | ones | | tens | ones | | tens | ones | | tens | ones |

	1	8
−	0	2

	7	7
−	3	1

	6	6
−	0	3

	4	4
−	1	0

	2	5
−	2	3

	5	6
−	4	2

	8	7
−	2	3

	3	3
−	2	1

	6	5
−	1	2

	8	2
−	1	1

	4	4
−	2	2

	5	6
−	1	0

	7	6
−	3	6

	8	6
−	4	5

	9	8
−	8	2

	4	6
−	1	2

	6	6
−	5	3

	7	7
−	4	2

	7	8
−	0	1

	0	7
−	0	1

	4	8
−	1	0

	5	5
−	0	3

	4	4
−	2	2

	5	6
−	5	2

	6	6
−	2	2

NAME: _____ DATE: _____

tens	ones		tens	ones		tens	ones		tens	ones		tens	ones

Row 1:
```
   tens | ones        tens | ones        tens | ones        tens | ones        tens | ones
    1  |  7            3  |  7            2  |  4            5  |  5            7  |  5
-   0  |  2        -   3  |  1        -   0  |  3        -   1  |  0        -   2  |  4
_____  _____  _____  _____  _____
```

Row 2:
```
    5  |  5            8  |  4            3  |  9            8  |  7            5  |  5
-   4  |  2        -   2  |  3        -   2  |  6        -   1  |  2        -   2  |  3
_____  _____  _____  _____  _____
```

Row 3:
```
    8  |  6            3  |  4            5  |  7            9  |  8            7  |  6
-   7  |  2        -   1  |  0        -   3  |  2        -   2  |  3        -   3  |  2
_____  _____  _____  _____  _____
```

Row 4:
```
    6  |  8            7  |  7            2  |  4            9  |  9            3  |  7
-   1  |  3        -   5  |  3        -   0  |  2        -   2  |  1        -   2  |  1
_____  _____  _____  _____  _____
```

Row 5:
```
    1  |  9            5  |  8            9  |  7            6  |  6            6  |  6
-   1  |  2        -   4  |  2        -   9  |  4        -   3  |  2        -   3  |  2
_____  _____  _____  _____  _____
```

38

NAME: _____ **DATE:** _____

	tens	ones
	1	7
−	0	2

	tens	ones
	4	7
−	3	1

	tens	ones
	2	4
−	0	3

	tens	ones
	5	4
−	1	0

	tens	ones
	2	9
−	2	6

	tens	ones
	5	5
−	4	2

	tens	ones
	3	4
−	2	3

	tens	ones
	3	7
−	2	6

	tens	ones
	8	7
−	1	2

	tens	ones
	4	6
−	1	3

	tens	ones
	5	5
−	3	2

	tens	ones
	6	6
−	1	0

	tens	ones
	3	7
−	3	7

	tens	ones
	9	9
−	4	5

	tens	ones
	8	7
−	4	2

	tens	ones
	5	7
−	1	3

	tens	ones
	9	9
−	5	5

	tens	ones
	5	5
−	4	2

	tens	ones
	7	7
−	2	1

	tens	ones
	6	8
−	1	1

	tens	ones
	9	7
−	9	2

	tens	ones
	8	5
−	5	2

	tens	ones
	5	8
−	4	2

	tens	ones
	9	7
−	5	2

	tens	ones
	9	8
−	6	2

39

NAME: _____ DATE: _____

| tens | ones | | tens | ones | | tens | ones | | tens | ones | | tens | ones |

	5	7
−	0	2

	3	6
−	3	1

	1	4
−	0	3

	5	6
−	1	0

	7	8
−	2	6

	5	5
−	4	2

	2	5
−	2	3

	3	9
−	2	6

	8	8
−	1	2

	5	5
−	1	3

	7	3
−	7	2

	3	4
−	1	0

	4	5
−	3	5

	6	6
−	4	5

	7	7
−	2	2

	6	7
−	1	2

	6	3
−	5	3

	5	4
−	4	2

	7	8
−	2	1

	9	7
−	9	1

	5	7
−	3	2

	6	8
−	2	2

	5	8
−	5	2

	6	7
−	4	2

	9	5
−	2	2

NAME: _____   **DATE:** _____

tens	ones		tens	ones		tens	ones		tens	ones		tens	ones

	1	7
+	0	1

	2	9
+	3	0

	2	4
+	5	3

	5	7
+	1	0

	5	3
+	2	6

	5	5
+	8	2

	1	5
+	5	3

	3	8
+	2	1

	7	6
+	1	2

	6	6
+	1	3

	1	6
+	6	2

	3	4
+	0	0

	2	2
+	1	7

	1	0
+	4	5

	1	7
+	9	2

	6	1
+	1	6

	1	3
+	6	4

	3	6
+	4	2

	7	3
+	2	3

	0	4
+	4	1

	5	7
+	5	2

	1	9
+	9	0

	1	0
+	9	0

	8	7
+	8	2

	5	7
+	4	1

41

NAME: _____ DATE: _____

	tens	ones		tens	ones		tens	ones		tens	ones		tens	ones
	1	7		2	7		2	4		5	4		2	3
+	0	2	+	3	1	+	0	3	+	1	0	+	2	6

	tens	ones		tens	ones		tens	ones		tens	ones		tens	ones
	5	5		1	5		3	3		8	6		4	3
+	4	2	+	2	3	+	2	6	+	1	2	+	2	3

	tens	ones		tens	ones		tens	ones		tens	ones		tens	ones
	1	4		4	4		3	2		1	2		6	7
+	7	2	+	1	0	+	3	7	+	6	5	+	8	2

	tens	ones		tens	ones		tens	ones		tens	ones		tens	ones
	6	7		2	3		6	6		7	7		0	8
+	3	2	+	5	2	+	4	2	+	2	1	+	9	1

	tens	ones		tens	ones		tens	ones		tens	ones		tens	ones
	5	6		1	7		8	6		7	4		0	2
+	4	2	+	1	2	+	9	2	+	2	3	+	9	2

NAME: _____   DATE: _____

tens	ones

	1	7
+	3	2

	2	7
+	4	2

	5	5
+	0	3

	5	4
+	3	0

	2	3
+	7	6

	4	5
+	6	2

	4	5
+	2	3

	4	3
+	2	6

	8	6
+	0	2

	4	4
+	2	3

	2	3
+	7	3

	6	4
+	2	2

	2	0
+	4	9

	2	3
+	6	5

	1	4
+	8	2

	6	7
+	2	2

	4	3
+	5	4

	3	5
+	5	2

	6	6
+	2	1

	2	8
+	6	1

	1	6
+	5	2

	1	5
+	8	2

	1	7
+	1	2

	1	9
+	9	0

	1	7
+	5	1

NAME: _____ DATE: _____

tens	ones
1	7
+ 6	2

tens	ones
2	7
+ 5	2

tens	ones
3	6
+ 5	3

tens	ones
5	4
+ 3	1

tens	one
2	0
+ 4	7

tens	ones
0	7
+ 4	2

tens	ones
5	8
+ 2	1

tens	ones
3	2
+ 5	6

tens	ones
8	3
+ 1	5

tens	ones
5	1
+ 1	6

tens	ones
2	9
+ 7	0

tens	ones
3	4
+ 5	4

tens	ones
5	2
+ 4	7

tens	ones
1	3
+ 6	5

tens	ones
2	6
+ 8	2

tens	ones
5	7
+ 5	2

tens	ones
2	4
+ 4	3

tens	ones
3	5
+ 7	2

tens	ones
7	8
+ 0	0

tens	ones
3	7
+ 9	2

tens	ones
3	3
+ 5	2

tens	ones
1	7
+ 9	2

tens	ones
1	2
+ 6	6

tens	ones
1	7
+ 7	2

tens	ones
1	5
+ 5	2

NAME: _____ DATE: _____

tens	ones		tens	ones		tens	ones		tens	ones		tens	ones
1	9		2	7		2	4		5	4		2	3
+ 0	0		+ 5	1		+ 0	1		+ 4	4		+ 6	6

6	6		7	5		5	7		8	6		6	6
+ 4	2		+ 2	3		+ 3	2		+ 1	3		+ 1	3

2	5		3	4		2	2		5	2		1	3
+ 7	2		+ 4	5		+ 4	6		+ 5	6		+ 5	2

6	7		1	3		3	4		7	8		0	7
+ 8	2		+ 5	5		+ 4	3		+ 3	1		+ 9	0

1	7		1	5		1	7		1	7		1	8
+ 9	2		+ 9	2		+ 7	1		+ 4	2		+ 8	1

45

NAME: _____ **DATE:** _____

tens	ones		tens	ones		tens	ones		tens	ones		tens	ones

	1	9			3	7			2	4			5	4			2	3
+	5	0		+	5	1		+	4	1		+	5	4		+	8	6

	5	7			9	8			5	7			8	5			6	6
+	4	2		+	2	1		+	4	2		+	2	3		+	3	1

	3	6			6	4			3	2			5	2			1	3
+	7	2		+	4	5		+	4	6		+	5	5		+	6	3

	6	7			1	3			3	4			7	8			0	5
+	1	2		+	6	6		+	5	5		+	4	1		+	9	1

	1	9			2	5			2	7			1	0			3	2
+	9	0		+	7	3		+	7	2		+	0	2		+	8	1

tens	ones

```
    2   8
+       3
_____
```

tens	ones

```
    9   3
+   4   4
_____
```

tens	ones

```
    2   0
+   3   0
_____
```

tens	ones

```
    9   2
+       2
_____
```

tens	ones

```
    5   9
+   2   2
_____
```

```
    6   1
+   1   4
_____
```

```
    4   6
+   2   2
_____
```

```
    1   5
+   1   6
_____
```

```
    3   2
+   1   2
_____
```

```
    4   4
+   4   0
_____
```

```
    5   7
+   5   2
_____
```

```
    4   4
+   1   8
_____
```

```
    7   0
+   2   1
_____
```

```
    8   8
+   1   4
_____
```

```
    3   1
+   2   2
_____
```

```
    4   8
+   4   6
_____
```

```
    2   7
+   2   9
_____
```

```
    6   9
+   5   5
_____
```

```
    2   8
+   3   8
_____
```

```
    9   5
+   2   5
_____
```

```
    8   7
+   3   3
_____
```

```
    3   8
+   1   3
_____
```

```
    2   3
+   3   2
_____
```

```
    4   9
+   4   1
_____
```

```
    3   1
+   8   8
_____
```

tens	ones

```
     4   6
 +   6   3
```

tens	ones

```
     7   2
 +   5   9
```

tens	ones

```
     1   4
 +   8   9
```

tens	ones

```
     2   1
 +   5   1
```

tens	ones

```
     7   0
 +   4   7
```

```
     8   5
 +   6   3
```

```
     6   8
 +   2   3
```

```
     6   9
 +   5   5
```

```
     2   5
 +   7   4
```

```
     1   5
 +   8   9
```

```
     2   4
 +   8   4
```

```
     3   6
 +   9   4
```

```
     7   6
 +   3   1
```

```
     2   8
 +   1   4
```

```
     6   1
 +   9   2
```

```
     7   8
 +   4   1
```

```
     9   7
 +   2   9
```

```
     4   9
 +   7   5
```

```
     6   8
 +   3   8
```

```
     9   5
 +   7   5
```

```
     5   7
 +   6   3
```

```
     5   8
 +   4   3
```

```
     4   3
 +   3   3
```

```
     8   9
 +   9   1
```

```
     5   1
 +   8   2
```

48

NAME: _____   DATE: _____

tens	ones		tens	ones		tens	ones		tens	ones		tens	ones

	5	6
+	6	5

	9	2
+	2	9

	3	4
+	8	9

	2	9
+	4	1

	6	0
+	5	7

	8	3
+	6	7

	2	8
+	2	3

	6	4
+	8	5

	5	5
+	5	4

	4	5
+	8	6

	2	9
+	6	4

	5	6
+	9	5

	7	2
+	8	1

	6	8
+	1	9

	3	1
+	9	6

	3	8
+	4	9

	5	7
+	8	9

	7	9
+	3	5

	9	8
+	3	2

	9	9
+	1	5

	1	7
+	8	3

	2	8
+	9	5

	4	1
+	6	1

	7	9
+	9	4

	5	5
+	6	4

49

tens	ones		tens	ones		tens	ones		tens	ones		tens	ones

Row 1

	7	7
+	6	9

	4	2
+	2	4

	6	7
+	7	9

	8	9
+	5	4

	6	8
+	6	7

Row 2

	5	4
+	5	7

	7	8
+	2	2

	4	4
+	9	5

	7	5
+	6	5

	3	5
+	8	9

Row 3

	5	4
+	4	7

	9	6
+	8	5

	9	1
+	5	2

	7	8
+	3	9

	7	1
+	3	9

Row 4

	8	8
+	7	3

	9	7
+	8	4

	3	4
+	7	5

	6	8
+	3	4

	7	5
+	4	5

Row 5

	8	7
+	4	3

	2	4
+	7	3

	7	6
+	4	2

	8	2
+	4	4

	5	9
+	1	4

NAME: _____ **DATE:** _____

tens	ones		tens	ones		tens	ones		tens	ones		tens	ones

	5	8
−	2	2

	6	7
−	4	1

	8	6
−	2	3

	4	6
−	1	9

	4	5
−	1	3

	8	6
−	7	2

	8	9
−	5	3

	9	3
−	9	1

	7	5
−	4	2

	8	2
−	6	1

	6	4
−	4	2

	7	6
−	7	0

	7	9
−	4	6

	9	6
−	5	5

	9	5
−	4	2

	7	6
−	2	2

	6	6
−	2	3

	4	7
−	3	2

	8	8
−	3	1

	6	7
−	4	1

	7	8
−	1	5

	9	5
−	4	3

	7	4
−	4	6

	9	6
−	1	2

	5	6
−	2	5

NAME: _____   DATE: _____

tens | ones	tens | ones	tens | ones	tens | ones	tens | ones
6 7	9 7	4 6	7 6	8 5
− 4 2	− 4 4	− 3 3	− 6 9	− 4 3

7 8	9 9	8 3	4 5	7 2
− 7 2	− 3 3	− 4 6	− 3 2	− 6 7

9 7	8 6	8 9	6 6	9 8
− 4 9	− 6 0	− 1 6	− 1 3	− 6 2

6 6	7 7	5 6	7 8	5 7
− 4 2	− 2 6	− 4 2	− 3 5	− 2 1

9 8	6 5	9 4	7 6	8 8
− 2 8	− 4 2	− 5 4	− 4 2	− 4 8

NAME: _____ **DATE:** _____

tens | ones	tens | ones	tens | ones	tens | ones	tens | ones

	tens	ones
	5	7
−	2	9

	tens	ones
	8	4
−	5	6

	tens	ones
	6	7
−	4	2

	tens	ones
	4	6
−	3	2

	tens	ones
	8	2
−	6	3

	tens	ones
	5	8
−	4	2

	tens	ones
	7	8
−	4	3

	tens	ones
	7	5
−	5	6

	tens	ones
	9	5
−	3	2

	tens	ones
	8	2
−	6	2

	tens	ones
	8	9
−	5	9

	tens	ones
	7	5
−	2	0

	tens	ones
	6	9
−	2	6

	tens	ones
	8	6
−	4	3

	tens	ones
	8	1
−	5	2

	tens	ones
	8	6
−	5	2

	tens	ones
	4	7
−	3	6

	tens	ones
	8	6
−	2	2

	tens	ones
	4	8
−	3	5

	tens	ones
	8	7
−	3	5

	tens	ones
	8	5
−	5	8

	tens	ones
	9	5
−	5	2

	tens	ones
	8	6
−	5	3

	tens	ones
	8	6
−	5	2

	tens	ones
	9	7
−	2	8

tens | ones	tens | ones	tens | ones	tens | ones	tens | ones

Row 1

```
   8  4        7  2        4  1        7  6        7  9
-  3  9     -  6  6     -  2  2     -  6  1     -  1  3
```

Row 2

```
   6  8        7  3        8  5        5  5        7  7
-  4  2     -  6  3     -  4  5     -  3  4     -  6  6
```

Row 3

```
   7  9        4  5        9  9        5  6        6  1
-  6  9     -  2  0     -  2  3     -  4  7     -  2  2
```

Row 4

```
   7  6        8  7        9  6        6  8        7  7
-  6  2     -  3  3     -  2  7     -  3  4        5  5
```

Row 5

```
   5  5        9  0        6  6        7  6        6  7
-  1  8     -  2  2     -  5  0     -  3  2     -  2  0
```

54

tens	ones		tens	ones		tens	ones		tens	ones		tens	ones

	7	4
−	3	7

	8	2
−	6	6

	9	1
−	2	2

	8	6
−	5	1

	8	9
−	2	3

	8	8
−	2	2

	7	3
−	3	3

	5	5
−	2	5

	9	5
−	5	4

	4	7
−	3	4

	8	9
−	6	0

	8	5
−	2	8

	7	9
−	2	2

	9	6
−	8	7

	8	1
−	2	7

	8	6
−	8	2

	8	0
−	3	0

	9	6
−	2	0

	7	8
−	3	0

	8	7
	6	5

	5	5
−	1	0

	9	0
−	0	2

	6	0
−	5	3

	7	6
−	1	2

	8	7
−	2	7

ANSWER SHEET

Row 1	19	58	27	64	49
Row 2	97	37	59	98	56
Row 3	85	44	59	57	99
Row 4	79	66	76	99	98
Row 5	49	79	89	46	39

Page 1

Row 1	47	27	38	84	69
Row 2	37	59	87	88	67
Row 3	95	34	69	89	85
Row 4	56	39	87	49	96
Row 5	68	26	95	19	48

Page 2

Row 1	45	78	47	54	49
Row 2	99	33	29	58	76
Row 3	65	42	53	97	9
Row 4	78	73	86	78	48
Row 5	109	116	67	125	39

Page 3

Row 1	45	88	55	74	34
Row 2	68	17	59	99	19
Row 3	25	46	52	67	79
Row 4	74	49	66	57	29
Row 5	86	36	88	59	19

Page 4

Row 1	69	57	37	65	29
Row 2	99	47	53	18	66
Row 3	105	74	55	47	98
Row 4	89	75	69	59	88
Row 5	89	49	44	112	129

Page 5

Row 1	15	16	41	44	24
Row 2	44	21	72	14	12
Row 3	37	36	45	53	35
Row 4	5	6	32	37	93
Row 5	25	35	33	37	13

Page 6

Row 1	25	16	21	44	22
Row 2	13	11	10	74	30
Row 3	24	24	20	21	14
Row 4	24	4	14	25	35
Row 5	56	14	54	6	74

Page 7

Row 1	15	15	21	34	30
Row 2	4	14	12	23	20
Row 3	22	44	20	20	16
Row 4	24	11	12	54	6
Row 5	15	5	51	35	15

Page 8

Row 1	23	13	14	45	11
Row 2	43	21	11	23	10
Row 3	1	34	11	12	5
Row 4	30	11	12	56	6
Row 5	2	5	14	33	50

Page 9

Row 1	17	5	21	44	1
Row 2	13	11	7	74	30
Row 3	1	24	2	10	5
Row 4	52	12	3	47	5
Row 5	4	25	4	4	2

Page 10

Row 1	28	57	27	54	69
Row 2	57	47	85	66	55
Row 3	35	43	47	55	67
Row 4	49	36	66	98	97
Row 5	98	188	109	108	80

Page 11

Row 1	29	59	17	62	89
Row 2	98	87	59	58	56
Row 3	85	44	69	117	99
Row 4	78	36	106	49	108
Row 5	68	100	195	105	89

Page 12

ANSWER SHEET

Row 1	39	58	24	84	39
Row 2	96	57	49	108	86
Row 3	95	47	69	54	69
Row 4	39	76	96	99	98
Row 5	54	29	46	69	108

Page 13

Row 1	59	59	27	66	49
Row 2	97	37	59	68	57
Row 3	88	94	89	57	39
Row 4	89	96	86	96	58
Row 5	106	49	139	108	107

Page 14

Row 1	38	69	49	67	79
Row 2	85	27	89	97	68
Row 3	89	47	69	48	68
Row 4	77	57	79	89	88
Row 5	66	77	47	69	69

Page 15

Row 1	35	7	3	40	3
Row 2	33	44	13	74	80
Row 3	21	22	10	11	14
Row 4	46	22	12	54	45
Row 5	13	24	22	35	64

Page 16

Row 1	66	35	23	43	10
Row 2	42	31	41	72	39
Row 3	13	36	20	12	25
Row 4	15	15	34	12	45
Row 5	34	26	16	55	56

Page 17

Row 1	13	4	63	46	42
Row 2	23	2	10	65	32
Row 3	4	55	51	2	2
Row 4	51	2	13	47	46
Row 5	7	54	32	25	87

Page 18

Row 1	96	56	74	26	0
Row 2	33	71	13	64	32
Row 3	3	20	51	32	17
Row 4	43	11	12	76	26
Row 5	41	15	5	5	4

Page 19

Row 1	10	36	41	45	11
Row 2	12	21	10	33	31
Row 3	31	22	64	23	65
Row 4	13	10	52	77	6
Row 5	16	17	23	13	15

Page 20

Row 1	17	87	39	76	78
Row 2	107	97	54	99	55
Row 3	149	87	99	86	179
Row 4	119	67	79	99	98
Row 5	59	119	89	145	107

Page 21

Row 1	59	69	77	66	49
Row 2	99	58	99	108	58
Row 3	77	48	58	67	29
Row 4	77	63	78	89	88
Row 5	109	19	108	59	99

Page 22

Row 1	89	51	37	66	49
Row 2	107	17	58	108	59
Row 3	65	64	99	87	89
Row 4	89	117	99	88	99
Row 5	19	97	77	90	81

Page 23

Row 1	17	118	87	67	29
Row 2	99	36	59	99	55
Row 3	158	45	60	57	109
Row 4	78	77	79	109	97
Row 5	106	34	109	69	68

Page 24

ANSWER SHEET

Row 1	19	58	27	64	49
Row 2	97	27	59	98	56
Row 3	85	44	59	57	99
Row 4	79	66	76	99	98
Row 5	68	147	67	108	48

Page 25

Row 1	15	26	11	35	22
Row 2	3	43	13	64	23
Row 3	53	25	31	11	5
Row 4	65	16	35	27	46
Row 5	23	14	34	22	43

Page 26

Row 1	14	16	51	24	1
Row 2	17	25	10	75	31
Row 3	21	44	20	21	5
Row 4	55	10	32	77	6
Row 5	24	6	54	76	25

Page 27

Row 1	55	13	33	46	10
Row 2	44	21	70	77	34
Row 3	13	74	55	37	15
Row 4	65	11	15	47	17
Row 5	34	2	24	33	4

Page 28

Row 1	13	35	22	47	41
Row 2	43	12	11	77	35
Row 3	4	25	50	10	45
Row 4	73	3	30	54	76
Row 5	5	15	93	14	35

Page 29

Row 1	15	56	25	24	0
Row 2	14	5	11	66	34
Row 3	6	34	22	21	14
Row 4	65	3	22	45	6
Row 5	7	75	6	51	13

Page 30

Row 1	68	58	77	66	69
Row 2	88	89	76	68	59
Row 3	165	74	79	97	69
Row 4	89	65	75	96	178
Row 5	67	98	148	137	96

Page 31

Row 1	69	128	67	75	59
Row 2	99	98	57	89	99
Row 3	98	47	58	59	79
Row 4	97	107	78	86	148
Row 5	58	48	69	19	95

Page 32

Row 1	18	87	29	68	89
Row 2	99	69	79	99	76
Row 3	105	48	55	98	39
Row 4	77	69	116	98	34
Row 5	87	160	120	188	138

Page 33

Row 1	59	69	57	68	53
Row 2	97	37	59	98	56
Row 3	85	44	59	57	99
Row 4	79	66	76	99	98
Row 5	108	149	118	69	12

Page 34

Row 1	39	89	79	66	79
Row 2	108	58	57	99	109
Row 3	97	88	29	98	115
Row 4	49	69	86	98	38
Row 5	88	129	77	108	109

Page 35

Row 1	54	26	91	54	70
Row 2	13	3	11	76	34
Row 3	11	12	22	23	15
Row 4	65	11	1	34	45
Row 5	84	53	4	24	54

Page 36

ANSWER SHEET

Row 1	16	46	63	34	2
Row 2	14	64	12	53	71
Row 3	22	46	40	41	16
Row 4	34	13	35	77	6
Row 5	38	52	22	4	44

Page 37

Row 1	15	6	21	45	51
Row 2	13	61	13	75	32
Row 3	14	24	25	75	44
Row 4	55	24	22	78	16
Row 5	7	16	3	34	34

Page 38

Row 1	15	16	21	44	3
Row 2	13	11	11	75	33
Row 3	23	56	0	54	45
Row 4	44	44	13	56	57
Row 5	5	33	16	45	36

Page 39

Row 1	55	5	11	46	52
Row 2	13	2	13	76	42
Row 3	1	24	10	21	55
Row 4	55	10	12	57	6
Row 5	25	46	6	25	73

Page 40

Row 1	18	59	77	67	79
Row 2	137	68	59	88	79
Row 3	78	34	39	55	109
Row 4	77	77	78	96	45
Row 5	109	109	100	169	98

Page 41

Row 1	19	58	27	64	49
Row 2	97	38	59	98	66
Row 3	86	54	69	77	149
Row 4	99	75	108	98	99
Row 5	98	29	178	97	94

Page 42

Row 1	49	69	58	84	99
Row 2	107	68	69	88	67
Row 3	96	86	69	88	96
Row 4	89	97	87	87	89
Row 5	68	97	29	109	68

Page 43

Row 1	79	79	89	85	67
Row 2	49	79	88	98	67
Row 3	99	88	99	78	108
Row 4	109	67	107	78	129
Row 5	85	109	78	89	67

Page 44

Row 1	19	78	25	98	89
Row 2	108	98	89	99	79
Row 3	97	79	68	108	65
Row 4	149	68	77	109	97
Row 5	109	107	88	59	99

Page 45

Row 1	69	88	65	108	109
Row 2	99	119	99	108	97
Row 3	108	109	78	107	76
Row 4	79	79	89	119	96
Row 5	109	98	99	12	113

Page 46

Row 1	31	137	50	94	81
Row 2	75	68	31	44	84
Row 3	109	62	91	102	53
Row 4	94	56	124	66	120
Row 5	120	51	55	90	119

Page 47

Row 1	109	131	103	72	117
Row 2	148	91	124	99	104
Row 3	108	130	107	42	153
Row 4	119	126	124	106	170
Row 5	120	101	76	180	133

Page 48

ANSWER SHEET

Row 1	121	121	123	70	117
Row 2	150	51	149	109	131
Row 3	93	151	153	87	127
Row 4	87	146	114	130	114
Row 5	100	123	102	173	119

Page 49

Row 1	146	66	146	143	135
Row 2	111	100	139	140	124
Row 3	101	181	143	117	110
Row 4	161	181	109	102	120
Row 5	130	97	118	126	73

Page 50

Row 1	36	26	63	27	32
Row 2	14	36	2	33	21
Row 3	22	6	33	41	53
Row 4	54	43	15	57	26
Row 5	63	52	28	84	31

Page 51

Row 1	25	53	13	7	42
Row 2	6	66	37	13	5
Row 3	48	26	73	53	36
Row 4	24	51	14	43	36
Row 5	70	23	40	34	40

Page 52

Row 1	28	28	25	14	19
Row 2	16	35	19	63	20
Row 3	30	55	43	43	29
Row 4	34	11	64	13	52
Row 5	27	43	33	34	69

Page 53

Row 1	45	6	19	15	66
Row 2	26	10	40	21	11
Row 3	10	25	76	9	39
Row 4	14	54	69	34	22
Row 5	27	68	16	44	47

Page 54

Row 1	37	16	69	35	66
Row 2	66	40	30	41	13
Row 3	29	57	57	9	54
Row 4	4	50	76	48	22
Row 5	45	88	7	74	60

Page 55

Thank You
For
Your
Purchase

Made in the USA
Monee, IL
24 March 2020

23852454R00037